Observations 3- common behaviors and natural principles

The goal of this book is to explain the common behaviors which seem like natural principles that I have observed over my lifetime. It explains, from multiple angles, to bring understanding of how these common behaviors and natural principles manifest. It also then provides examples on how to apply them to one's daily life for success, happiness, and financial well-being. It should also provide the knowledge "why" one is not satisfied with life or things are not going as intended.

Success, happiness, and financial well-being follows a recipe, a process and this book explains the knowledge to achieve them.

All professionals make things seem easy because they have mastered the knowledge and process of their profession.

The objective then is after understanding these common behaviors which are like natural principles, the workings and manifestations, and the process for you to join the successful happy people who also have financial well-being.

Part 1

Basic explanation of the observations

Part 2

Graphic examples of the principles and observations

Part 3

Further explanation with detailed examples

Part 4

Applications for your success

Part 5

Further research and observations

Part 1

Basic explanation of the observations

All the principles of the observations focus on one principle such as equilibrium on multiple areas with examples of each. Below is the outline of how part 3 explains each from the multiple angles.

Format though out part 3.

1 equals 1 (Net Zero Law 1-1=0 equilibrium)

detailed explanation of Equilibrium with examples

detailed explanation of Equilibrium in nature with examples

detailed explanation of Equilibrium in the economy with examples

detailed explanation of Equilibrium in the physical science with examples

detailed explanation of Equilibrium in social science with examples

detailed explanation of Equilibrium in human nature with examples

detailed explanation of Equilibrium in astronomy with examples detailed explanation of Equilibrium in physics with examples

detailed explanation of Equilibrium with physical forces with examples

detailed explanation of Equilibrium in electronics with examples

applications of Equilibrium in daily life

applications of Equilibrium in daily life for happiness

applications of Equilibrium in daily life for success

applications of Equilibrium in daily life in relationships

applications of Equilibrium in daily life for financial balance

applications of Equilibrium in daily life for work life balance

applications of Equilibrium in goal setting

How to achieve mental peace

The observations discussed are:

Equilibrium

Ripple effect

Butterfly effect

Multiplier effect

Domino effect

Cascading effect

Chain reaction effect

Snowball effect

Duality in nature

Natural state return to equilibrium

Detailed explanation of the pendulum with examples

Detailed explanation of cycles with examples

Level theory

   Infinite dimensions within infinite worlds

Law of depletive returns

Maxima

Minima

Feast or famine

Upper and lower limits with equilibrium

The more it approaches equilibrium the more unstable it can become

Natural variation keeps stability

Lewin force field analysis

Path of least resistance

The hump of resistance to make a change

Law of attraction

Bullwhip

Entropy

Self organizing system

Chaos

Give to get

Reap what you sow

And much much more to plant the seed how underlying principles manifest in so many different forms and the parts follow these principles like an apple falling from a tree obeys the principle of gravity.

Part 2

Graphic examples of the principles and observations

Basic graphics depicting the levels and regions of energy

Region ID		
Above	Infinite	Infinite resistance
A	3	
B	2	
C	1	Variation but stable region
Center	0	Total equilibrium-Net zero, High risk of becoming unstable
D	-1	Variation but stable region
E	-2	
F	-3	
Below	-infinite	Infinite resistance

Identification of the energy force field with a push and pull towards equilibrium.

Region ID	Force field	
Above	Infinite	Infinite resistance
A	3	
B	2	
C	1	Variation but stable region
Center	0	Total equilibrium-Net zero, High risk of becoming unstable
D	-1	Variation but stable region
E	-2	
F	-3	
Below	-infinite	Infinite resistance

Levels explained and the multi-infinite dimensions that are multi-infinitive

Region ID	Force field	
Above	Infinite	Infinite resistance
A	3	
B	2	
C	1	Variation but stable region
Center	0	Total equilibrium-Net zero, High risk of becoming unstable
D	-1	Variation but stable region
E	-2	
F	-3	
Below	-infinite	Infinite resistance

There are levels. These are infinite in infinite dimensions
The Mandelbrot is an example of this infinite multidimension.
A comparison is the plane Earth to a grain of sand to the atom

Magnitudes of the push and pull to equilibrium

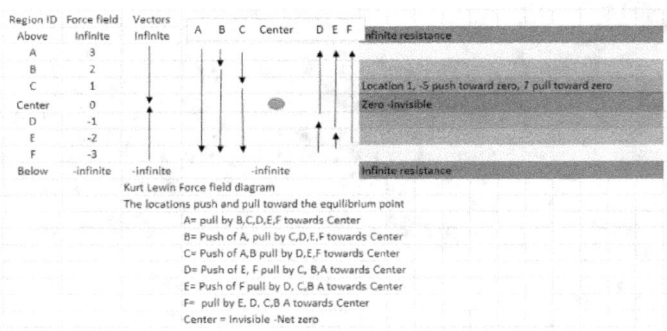

Kurt Lewin Force field diagram
The locations push and pull toward the equilibrium point
A= pull by B,C,D,E,F towards Center
B= Push of A, pull by C,D,E,F towards Center
C= Push of A,B pull by D,E,F towards Center
D= Push of E, F pull by C, B,A towards Center
E= Push of F pull by D, C,B A towards Center
F= pull by E, D, C,B A towards Center
Center = Invisible -Net zero

Duality is required to exist away from equilibrium

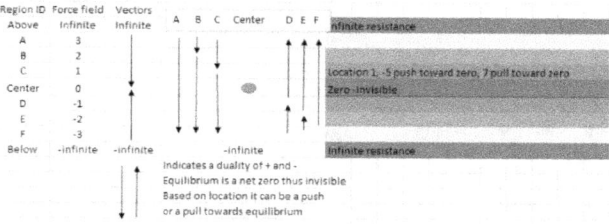

A normal stable cycle

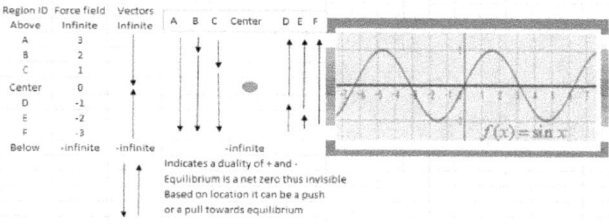

Chaos from a system at equilibrium do to not having the stability of cycling

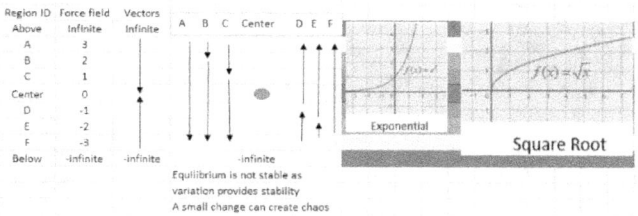

Part 3

Further explanation with detailed examples

Equilibrium refers to a state in which the supply of a product or service is balanced with the demand for it, resulting in a stable price. At equilibrium, the forces of supply and demand are in balance, and there is no incentive for the price to change.

Here are a few examples to help illustrate the concept of equilibrium:

1. The market for apples: Suppose the price of apples is $1 per pound. At this price, farmers are willing to sell a certain quantity of apples, and consumers are willing to buy a certain quantity. If the quantity of apples that farmers are willing to sell is equal to the quantity that consumers are willing to buy, then the market is in equilibrium. If the price were higher, then there would be a surplus of apples because farmers would be willing to sell more than consumers are willing to buy. If the price were lower, then there would be a shortage of apples because consumers would be willing to buy more than farmers are willing to sell.

2. The market for labor: In the labor market, the supply of labor represents the number of people looking for work, and the demand for labor represents the number of jobs available. At equilibrium, the number of people looking for work is equal to the number of jobs available. If there are more jobs available than people looking for work, then the market is in surplus and wages may increase as employers

compete to attract workers. If there are more people looking for work than there are jobs available, then the market is in shortage and wages may decrease as workers compete for fewer job openings.

3. The market for a particular stock: In the stock market, the supply of a particular stock represents the number of shares available for purchase, and the demand for the stock represents the number of people willing to buy it. At equilibrium, the number of shares available for purchase is equal to the number of people willing to buy it, resulting in a stable stock price. If there is more demand for the stock than there are shares available for purchase, then the price of the stock may increase. If there is more supply than demand, then the price may decrease.

In economics, equilibrium refers to a state in which the supply of a product or service is balanced with the demand for it, resulting in a stable price. At equilibrium, the forces of supply and demand are in balance, and there is no incentive for the price to change.

Here are a few examples of equilibrium in the economy:

1. The market for a particular good or service: In a competitive market, the price of a good or service will tend to adjust until it reaches a level at which the quantity of the good or service that producers are willing to supply is equal to the quantity that consumers are willing to demand. For example, suppose the market for apples is in equilibrium at a price of $1 per pound. At this price, farmers are willing to sell a certain quantity of apples, and consumers are willing to buy a certain quantity. If the price were higher, then there would be a surplus of apples because farmers would be willing to sell more than consumers are willing to buy. If the price were lower, then there would be a shortage of apples because consumers would be willing to buy more than farmers are willing to sell.

2. The labor market: In the labor market, the supply of labor represents the number of people looking for work, and the demand for labor represents the number of jobs available. At equilibrium, the number of people looking for work is equal to the number of jobs

available. If there are more jobs available than people looking for work, then the market is in surplus and wages may increase as employers compete to attract workers. If there are more people looking for work than there are jobs available, then the market is in shortage and wages may decrease as workers compete for fewer job openings.

3. The stock market: In the stock market, the supply of a particular stock represents the number of shares available for purchase, and the demand for the stock represents the number of people willing to buy it. At equilibrium, the number of shares available for purchase is equal to the number of people willing to buy it, resulting in a stable stock price. If there is more demand for the stock than there are shares available for purchase, then the price of the stock may increase. If there is more supply than demand, then the price may decrease.

In nature, equilibrium refers to a state in which all the factors influencing a particular system are balanced and in stable proportions. In other words, there is a balance between the various elements of the system, and the system tends to remain stable over time.

Here are a few examples of equilibrium in nature:

1. Population dynamics: In population dynamics, equilibrium is achieved when the birth rate is equal to the death rate, resulting in a stable population size. For example, if a species of rabbits has a birth rate of 5 rabbits per month and a death rate of 5 rabbits per month, then the population will remain stable at a particular size. If the birth rate exceeds the death rate, then the population will grow, while if the death rate exceeds the birth rate, then the population will decline.

2. Ecological balance: In an ecosystem, equilibrium is achieved when the various species present are in balance with one another and with their environment. For example, in a forest ecosystem, there may be a balance between the number of trees, the number of herbivores (such as deer) that feed on the trees, and the number of carnivores (such as wolves) that prey on the herbivores. If one species becomes too numerous, it may disrupt the balance of the ecosystem, potentially leading to the decline of other species.

3. Chemical reactions: In chemistry, equilibrium is achieved when the rates of the forward and reverse reactions are equal, resulting in a stable concentration of reactants and products. For example, in the reaction between hydrogen gas and oxygen gas to form water, equilibrium is achieved when the rate at which hydrogen and oxygen molecules combine to form water is equal to the rate at which water molecules break down into hydrogen and oxygen.

In physical science, equilibrium refers to a state in which a system is balanced and its properties do not change over time. This can occur when the forces acting on a system are balanced and the system is not being subjected to any external forces or influences.

Here are a few examples of equilibrium in physical science:

1. Thermal equilibrium: In thermodynamics, thermal equilibrium is achieved when two objects at different temperatures are placed in thermal contact with each other. After a sufficient amount of time, the two objects will reach the same temperature, at which point they will be in thermal equilibrium. For example, if a cup of hot water and a cup of cold water are placed in contact with each other, the hot water will transfer some of its heat to the cold water until both cups reach the same temperature.

2. Mechanical equilibrium: In mechanics, a system is in mechanical equilibrium when the forces acting on it are balanced and the system is not accelerating. For example, a book sitting on a table is in mechanical equilibrium because the force of gravity pulling it downward is balanced by the force of the table pushing upward.

3. Electrostatic equilibrium: In electrostatics, a system is in electrostatic equilibrium when the electric charges within the system are balanced and there is no net electric force acting on the

system. For example, if a positively charged object and a negatively charged object are placed near each other, the positive and negative charges will redistribute themselves until the electric forces within the system are balanced.

In social science, equilibrium refers to a state in which a system is balanced and its properties do not change over time. This can occur when the various elements within a system are in balance and there are no external forces or influences acting on the system.

Here are a few examples of equilibrium in social science:

1. Game theory: In game theory, equilibrium refers to a strategy or set of strategies that are stable and cannot be improved upon by any individual player. For example, in the "prisoner's dilemma" game, the equilibrium strategy for both players is to confess, because neither player can improve their outcome by changing their strategy.

2. Political science: In political science, equilibrium can refer to a balance of power among different countries or groups, such as the balance of power among different branches of government within a single country. For example, the separation of powers between the executive, legislative, and judicial branches of government in the United States is designed to achieve a balance of power and prevent any one branch from becoming too dominant.

3. Sociology: In sociology, equilibrium can refer to a balance or stability within a social system, such as a balance between different social groups or a balance between individual and group interests. For example, if there is a

balance of power between different social classes in a society, then the society may be more stable and less prone to conflict.

It is important to note that the concept of equilibrium is primarily used in the natural and social sciences to describe the balance or stability of systems. In the context of human nature, the concept of equilibrium may not be as applicable. However, it is possible to interpret the concept of equilibrium in a more general sense as a balance or stability within an individual or society.

Here are a few examples of how the concept of equilibrium might be applied to human nature:

1. Emotional equilibrium: In psychology, emotional equilibrium refers to a balance or stability in an individual's emotional state. For example, an individual who is able to regulate their emotions and maintain a sense of balance in their emotional life may be said to have emotional equilibrium.

2. Social equilibrium: In sociology, social equilibrium can refer to a balance or stability within a social system, such as a balance between different social groups or a balance between individual and group interests. For example, a society in which there is a balance of power between different social classes may be more stable and less prone to conflict.

3. Moral equilibrium: In philosophy, the concept of moral equilibrium might refer to a balance or stability in an individual's moral beliefs and values. For example, an individual who is able to maintain a consistent set of moral beliefs and

values in the face of changing circumstances may be said to have moral equilibrium.

In astronomy, the concept of equilibrium can be applied to celestial objects and systems. For example:

1. Hydrostatic equilibrium: A celestial object is in hydrostatic equilibrium when the gravitational force acting on it is balanced by the pressure within the object. For example, a star is in hydrostatic equilibrium when the force of gravity pulling inward is balanced by the pressure of the gas and plasma within the star pushing outward.

2. Thermal equilibrium: A celestial object is in thermal equilibrium when the energy it absorbs from its surroundings is equal to the energy it radiates into space. For example, a planet is in thermal equilibrium when the energy it absorbs from the sun is equal to the energy it radiates back into space as heat.

3. Orbital equilibrium: A celestial object in orbit around another object is in orbital equilibrium when the gravitational force of the larger object is balanced by the centripetal force of the smaller object. For example, the Earth is in orbital equilibrium around the sun, with the gravitational force of the sun pulling the Earth inward balanced by the centripetal force of the Earth's motion keeping it in its orbit.

In physics, equilibrium can refer to a state in which the forces acting on a system are balanced and the system is not accelerating. When a system is in equilibrium, the net force acting on it is zero, and the system will not change its state of motion.

Here are a few examples of equilibrium involving physical forces:

1. Mechanical equilibrium: In mechanics, a system is in mechanical equilibrium when the forces acting on it are balanced and the system is not accelerating. For example, a book sitting on a table is in mechanical equilibrium because the force of gravity pulling it downward is balanced by the force of the table pushing upward.

2. Electrostatic equilibrium: In electrostatics, a system is in electrostatic equilibrium when the electric charges within the system are balanced and there is no net electric force acting on the system. For example, if a positively charged object and a negatively charged object are placed near each other, the positive and negative charges will redistribute themselves until the electric forces within the system are balanced.

3. Fluid equilibrium: In fluid dynamics, a fluid is in equilibrium when the forces acting on it are balanced and the fluid is not accelerating. For example, if a container of water is filled to the brim and not moving, the force of gravity pulling the water downward is balanced by the

force of the container pushing upward, resulting in equilibrium.

In physics, equilibrium refers to a state in which a system is balanced and its properties do not change over time. This can occur when the forces acting on a system are balanced and the system is not being subjected to any external forces or influences.

Here are a few examples of equilibrium in physics:

1. Thermal equilibrium: In thermodynamics, thermal equilibrium is achieved when two objects at different temperatures are placed in thermal contact with each other. After a sufficient amount of time, the two objects will reach the same temperature, at which point they will be in thermal equilibrium. For example, if a cup of hot water and a cup of cold water are placed in contact with each other, the hot water will transfer some of its heat to the cold water until both cups reach the same temperature.

2. Mechanical equilibrium: In mechanics, a system is in mechanical equilibrium when the forces acting on it are balanced and the system is not accelerating. For example, a book sitting on a table is in mechanical equilibrium because the force of gravity pulling it downward is balanced by the force of the table pushing upward.

3. Electrostatic equilibrium: In electrostatics, a system is in electrostatic equilibrium when the electric charges within the system are balanced and there is no net electric force acting on the system. For example, if a positively charged

object and a negatively charged object are placed near each other, the positive and negative charges will redistribute themselves until the electric forces within the system are balanced.

In electronics, equilibrium can refer to a state in which a system is balanced and its properties do not change over time. This can occur when the various elements within a system are in balance and there are no external forces or influences acting on the system.

Here are a few examples of equilibrium in electronics:

1. Electrical equilibrium: In electrical engineering, a system is in electrical equilibrium when the electric charges within the system are balanced and there is no net electric force acting on the system. For example, if a circuit is in equilibrium, the electric current flowing through the circuit will be constant and the voltage across the circuit will not change.

2. Thermal equilibrium: In electronic devices, thermal equilibrium can be achieved when the heat generated by the device is equal to the heat dissipated into the surrounding environment. For example, a computer processor will reach thermal equilibrium when the heat generated by the processor is equal to the heat dissipated through the heat sink and into the air.

3. Magnetic equilibrium: In electromagnetism, a system is in magnetic equilibrium when the magnetic forces within the system are balanced and there is no net magnetic force acting on the system. For example, a bar magnet in equilibrium will not experience any net force and will not accelerate.

There are many ways in which the concept of equilibrium is applicable to daily life. Here are a few examples:

1. Temperature regulation: The human body is constantly seeking thermal equilibrium with its environment. When the body is too hot, it will perspire and lose heat through evaporation in order to cool down. When the body is too cold, it will shiver and generate heat in order to warm up.

2. Financial markets: In financial markets, the concept of equilibrium is used to describe the balance between supply and demand for a particular asset, such as a stock or a currency. The price of the asset will tend to adjust until it reaches a level at which the quantity of the asset that buyers are willing to purchase is equal to the quantity that sellers are willing to sell.

3. Chemical reactions: The concept of equilibrium is important in chemistry, as it describes the balance between the rates of forward and reverse reactions. For example, the process of digestion in the human body is a series of chemical reactions that are in equilibrium.

4. Construction: In construction, the concept of equilibrium is used to ensure that the forces acting on a structure are balanced and the

structure is stable. For example, in the design of a bridge, engineers must ensure that the forces of the weight of the bridge and the loads placed on the bridge are balanced so that the bridge does not collapse.

Here are a few ways in which the concept of equilibrium can be applied to daily life in order to promote happiness:

1. Emotional equilibrium: Maintaining emotional equilibrium can be important for happiness. This can involve finding balance in one's emotional life and learning to regulate one's emotions in a healthy way. This may involve finding healthy ways to cope with stress, such as through exercise, meditation, or therapy.

2. Work-life balance: Achieving balance between one's work and personal life can be important for happiness. This may involve setting boundaries and making time for activities and relationships outside of work.

3. Physical balance: Maintaining physical balance and stability can help promote happiness by reducing the risk of falls and injuries, and by promoting physical health and well-being. This may involve engaging in regular physical activity, practicing good posture, and taking care of one's physical health.

4. Social balance: Maintaining a balance between solitude and social interaction can be important for happiness. This may involve finding a balance between time spent alone and time spent with others, and cultivating meaningful relationships.

Here are a few ways in which the concept of equilibrium can be applied to daily life in order to achieve success:

1. Time management: Achieving balance and equilibrium in how one manages their time can be important for success. This may involve setting goals and prioritizing tasks, and finding a balance between work and leisure.

2. Physical health: Maintaining physical balance and stability can be important for success, as good physical health can help one to be more productive and efficient. This may involve engaging in regular physical activity, getting enough sleep, and eating a healthy diet.

3. Emotional balance: Maintaining emotional equilibrium can help one to be more focused and productive, and can also improve relationships with others. This may involve finding healthy ways to cope with stress, such as through exercise, meditation, or therapy.

4. Social balance: Maintaining a balance between solitude and social interaction can be important for success, as social connections can provide support and opportunities for growth. This may involve cultivating meaningful relationships and finding a balance between time spent alone and time spent with others.

Here are a few ways in which the concept of equilibrium can be applied to daily life in relationships:

1. Communication: Maintaining balance in communication can be important for successful relationships. This may involve finding a balance between speaking and listening, and being open and honest with one's communication.

2. Boundaries: Setting and maintaining boundaries can help to achieve equilibrium in relationships, as it allows individuals to have their own space and autonomy while still being connected to others.

3. Emotional balance: Maintaining emotional equilibrium can help to create a sense of stability and balance in relationships. This may involve finding healthy ways to cope with stress, such as through exercise, meditation, or therapy, and learning to regulate one's emotions in a healthy way.

4. Physical balance: Maintaining physical balance and stability can be important for relationships, as it can help to reduce the risk of falls and injuries, and promote physical health and well-being. This may involve engaging in regular physical activity, practicing good posture, and taking care of one's physical health.

Here are a few ways in which the concept of equilibrium can be applied to daily life in order to achieve financial balance:

1. Budgeting: Creating and sticking to a budget can help to achieve financial equilibrium by balancing income and expenses. This can involve setting financial goals, tracking spending, and making adjustments as needed.

2. Debt management: Maintaining a balance between debt and assets can be important for financial stability. This may involve paying off high-interest debt, such as credit card debt, and making a plan to pay off other debts in a timely manner.

3. Savings: Achieving balance between saving and spending can be important for financial stability. This may involve setting aside a portion of income for savings and emergencies, while also leaving room for enjoyable expenses.

4. Investment: Finding a balance between risk and reward can be important in investment decisions. This may involve diversifying one's investment portfolio in order to balance risk and potential returns.

Here are a few ways in which the concept of equilibrium can be applied to daily life in order to achieve work-life balance:

1. Time management: Managing one's time effectively can help to balance the demands of work and personal life. This may involve setting priorities, creating a schedule, and finding ways to be more efficient.

2. Boundaries: Setting boundaries between work and personal life can help to achieve balance. This may involve setting limits on the amount of time spent working, and making time for activities and relationships outside of work.

3. Work-life integration: Finding ways to integrate work and personal life can also help to achieve balance. This may involve finding a job that allows for flexibility, such as working remotely or having a more flexible schedule.

4. Work-life separation: Maintaining separation between work and personal life can also be important for balance. This may involve finding ways to disconnect from work when not on the job, such as by turning off work-related notifications outside of work hours.

Here are a few ways in which the concept of equilibrium can be applied to goal setting:

1. Balancing short-term and long-term goals: Achieving balance between short-term and long-term goals can help to ensure that progress is made in the present while also keeping long-term aspirations in mind.

2. Balancing achievable and challenging goals: Finding a balance between goals that are achievable and those that are more challenging can help to motivate and inspire progress.

3. Balancing personal and professional goals: Maintaining balance between personal and professional goals can help to ensure that progress is made in all areas of one's life.

4. Balancing solo and collaborative goals: Finding a balance between solo and collaborative goals can help to ensure that progress is made both independently and as part of a team.

Here are a few tips for achieving mental peace:

1. Practice mindfulness: Being mindful involves paying attention to the present moment without judgment. This can help to bring a sense of calm and clarity to the mind.

2. Engage in relaxation techniques: There are many relaxation techniques that can help to bring a sense of peace to the mind, such as deep breathing, meditation, or progressive muscle relaxation.

3. Exercise regularly: Physical activity has been shown to have a positive effect on mental well-being. Engaging in regular exercise can help to reduce stress and improve mood.

4. Get enough sleep: Adequate sleep is important for mental well-being. Aim for 7-9 hours of sleep per night to help promote mental peace.

5. Cultivate positive relationships: Strong social connections have been shown to have a positive effect on mental well-being. Cultivating positive relationships with friends and loved ones can help to bring a sense of peace and support to one's life.

The "ripple effect" refers to the way in which one event can have a series of consequences or impacts on other related events. It is often used to describe the way in which a small change can have a significant impact on a larger system.

Here are a few examples of the ripple effect in action:

1. A change in government policy can have a ripple effect on the economy. For example, if a government increases taxes on certain goods, it may lead to a decrease in demand for those goods, which in turn could lead to job losses in the industries that produce those goods. This could then lead to a decrease in overall consumer spending, which could have further consequences for the economy.

2. A natural disaster can have a ripple effect on the people and communities affected by it. For example, a hurricane may destroy homes and infrastructure, which could lead to a shortage of housing and disrupt people's livelihoods. This could then have a ripple effect on the local economy, as people may have less money to spend on goods and services.

3. A change in technology can have a ripple effect on various industries. For example, the rise of streaming services has had a significant impact on the entertainment industry, as it has led to a decline in demand for traditional forms of media such as DVDs and CDs. This has had a ripple effect on the production and distribution

of these media, as well as on the people who work in these industries.

4. A personal decision can have a ripple effect on one's social circle. For example, if someone decides to end a relationship, it may have a ripple effect on their mutual friends, who may have to choose sides or may be saddened by the breakup. This could lead to changes in the dynamic of the social circle and potentially even the loss of friendships.

The ripple effect can also be observed in nature, where one event can have a series of consequences or impacts on other related events. Here are a few examples of the ripple effect in nature:

1. A change in the population size of a species can have a ripple effect on the ecosystem. For example, if the population of predators decreases, it could lead to an increase in the population of their prey. This could then have a ripple effect on the plants that the prey species feeds on, as the increased grazing could lead to a decrease in plant population.

2. A change in the temperature of the ocean can have a ripple effect on marine life. For example, warming ocean temperatures have been linked to the bleaching of coral reefs, which can have a ripple effect on the entire ecosystem. Coral reefs provide habitat for a diverse range of marine life, and their loss can have consequences for the species that rely on them.

3. A change in the quality of the air can have a ripple effect on plants and animals. For example, an increase in air pollution can lead to a decrease in the health of plants, which can then have a ripple effect on the animals that rely on those plants for food.

4. A change in the availability of water can have a ripple effect on the ecosystems that depend on it. For example, a drought can lead to a decrease in the availability of water for plants,

which can then have a ripple effect on the animals that rely on those plants for food and habitat.

The ripple effect in the economy refers to the way in which one event or change can have a series of consequences or impacts on other related events or sectors. Here are a few examples of the ripple effect in the economy:

1. A change in interest rates can have a ripple effect on the economy. For example, if the central bank increases interest rates, it can lead to a decrease in borrowing, as loans become more expensive. This can then have a ripple effect on consumer spending and investment, as people may have less money to spend and businesses may be less likely to invest in new projects.

2. A change in the value of a currency can have a ripple effect on the economy. For example, if the value of a country's currency decreases, it can make exports cheaper and more competitive on the global market. This could lead to an increase in demand for the country's exports, which could have a ripple effect on the domestic economy.

3. A change in government policy can have a ripple effect on the economy. For example, if a government increases taxes on certain goods or services, it can lead to a decrease in demand for those goods and services. This can then have a ripple effect on the industries that produce those goods and services, as well as on the people who work in those industries.

4. A change in the price of a commodity can have a ripple effect on the economy. For example, if the price of oil increases, it can lead to higher costs for businesses and consumers, which can then have a ripple effect on inflation and overall economic activity.

The ripple effect in physical science refers to the way in which one event or change can have a series of consequences or impacts on other related events or phenomena. Here are a few examples of the ripple effect in physical science:

1. A change in the velocity of an object can have a ripple effect on its motion. For example, if you throw a ball at a certain speed and angle, it will follow a specific trajectory. If you then increase the speed of the ball, it will follow a different trajectory. This change in velocity can have a ripple effect on the distance and height reached by the ball.

2. A change in the temperature of a substance can have a ripple effect on its properties. For example, if you heat up a solid, it will eventually reach its melting point and turn into a liquid. This change in temperature can have a ripple effect on the substance's density, viscosity, and other physical properties.

3. A change in the pressure of a gas can have a ripple effect on its volume. According to the ideal gas law, the pressure, volume, and temperature of a gas are inversely related. This means that if you increase the pressure of a gas, its volume will decrease. This change in pressure can have a ripple effect on the density and concentration of the gas.

4. A change in the concentration of a solution can have a ripple effect on its properties. For

example, if you increase the concentration of a solute in a solvent, it can affect the freezing point, boiling point, and other physical properties of the solution.

The ripple effect in social science refers to the way in which one event or change can have a series of consequences or impacts on other related events or phenomena in society. Here are a few examples of the ripple effect in social science:

1. A change in social norms can have a ripple effect on behavior. For example, if a society starts to view a certain behavior as unacceptable, it can lead to a decrease in the prevalence of that behavior. This change in social norms can have a ripple effect on the attitudes and values of the people in that society.

2. A change in technology can have a ripple effect on society. For example, the widespread adoption of smartphones has had a significant impact on the way people communicate and access information. This change in technology has had a ripple effect on various industries, such as the media and telecommunications industries, as well as on the way people work and interact with each other.

3. A change in the political climate can have a ripple effect on society. For example, if a country becomes more authoritarian, it can lead to a decrease in civil liberties and an increase in government control. This change in the political climate can have a ripple effect on the way people live their lives and express their opinions.

4. A change in economic conditions can have a ripple effect on society. For example, if a country experiences an economic recession, it can lead to a decrease in employment and a decrease in consumer spending. This change in economic conditions can have a ripple effect on the well-being of the people in that society, as well as on the stability of the government.

The ripple effect in human nature refers to the way in which one person's actions or decisions can have a series of consequences or impacts on others. Here are a few examples of the ripple effect in human nature:

1. A change in one person's behavior can have a ripple effect on the behavior of those around them. For example, if a person starts to exercise regularly, it could inspire others to do the same. This change in behavior could have a ripple effect on the overall health and well-being of the group.

2. A change in one person's beliefs or values can have a ripple effect on their relationships. For example, if a person starts to prioritize environmental sustainability, they may choose to live a more eco-friendly lifestyle, which could influence the choices of those around them. This change in beliefs could have a ripple effect on the behavior and values of the group.

3. A change in one person's financial circumstances can have a ripple effect on their relationships and their ability to contribute to their community. For example, if a person experiences a financial windfall, they may be able to support their loved ones or contribute to charitable causes. This change in financial circumstances could have a ripple effect on the well-being of the group.

4. A change in one person's attitude or outlook can have a ripple effect on their relationships

and their overall well-being. For example, if a person starts to adopt a more positive attitude, it could lead to improved relationships and a better quality of life. This change in attitude could have a ripple effect on the well-being of the group.

The ripple effect in astronomy refers to the way in which one celestial event or change can have a series of consequences or impacts on other related events or phenomena in the universe. Here are a few examples of the ripple effect in astronomy:

1. A supernova explosion can have a ripple effect on the surrounding galaxies and celestial bodies. A supernova is a massive explosion that occurs when a star collapses and explodes. The shockwaves from a supernova can have a ripple effect on the gases and dust in the surrounding area, which can then lead to the formation of new stars and planets.

2. A change in the mass of a celestial body can have a ripple effect on its orbit. For example, if a planet gains mass, it will have a stronger gravitational pull, which could cause its orbit to change. This change in mass could have a ripple effect on the orbits of other celestial bodies in the system.

3. A change in the temperature of a celestial body can have a ripple effect on its atmosphere and surface. For example, if a planet's temperature increases, it could lead to a change in the atmospheric composition and the presence of water on the surface. This change in temperature could have a ripple effect on the habitability of the planet.

4. A change in the activity of a celestial body can have a ripple effect on its surroundings. For

example, if a star increases its activity, it could lead to a change in the radiation levels in the surrounding area. This change in activity could have a ripple effect on the atmospheres and surfaces of nearby planets.

The ripple effect in physics refers to the way in which one physical event or change can have a series of consequences or impacts on other related events or phenomena. Here are a few examples of the ripple effect in physics:

1. A change in the velocity of an object can have a ripple effect on its momentum. Momentum is the product of an object's mass and velocity. If you increase the velocity of an object, you will also increase its momentum. This change in velocity can have a ripple effect on the object's ability to do work or cause a change in its environment.

2. A change in the temperature of a substance can have a ripple effect on its state. If you heat up a solid, it will eventually reach its melting point and turn into a liquid. This change in temperature can have a ripple effect on the substance's physical properties, such as its density and viscosity.

3. A change in the pressure of a gas can have a ripple effect on its volume. According to the ideal gas law, the pressure, volume, and temperature of a gas are inversely related. This means that if you increase the pressure of a gas, its volume will decrease. This change in pressure can have a ripple effect on the density and concentration of the gas.

4. A change in the position of an object can have a ripple effect on its potential energy. Potential

energy is the energy an object possesses due to its position or configuration. If you change the position of an object, you will also change its potential energy. This change in position can have a ripple effect on the object's ability to do work or cause a change in its environment.

The ripple effect with physical forces refers to the way in which one physical force can have a series of consequences or impacts on other related forces or phenomena. Here are a few examples of the ripple effect with physical forces:

1. A change in the magnitude of a force can have a ripple effect on the motion of an object. If you apply a greater force to an object, it will accelerate more. This change in force can have a ripple effect on the object's velocity and momentum.

2. A change in the direction of a force can have a ripple effect on the motion of an object. If you change the direction in which you apply a force to an object, it will change the object's direction of motion. This change in force direction can have a ripple effect on the object's velocity and momentum.

3. The presence of a force can have a ripple effect on the equilibrium of a system. Equilibrium refers to the state of balance in a system. The presence of a force can disturb the equilibrium of a system and cause a change in its motion. This change in equilibrium can have a ripple effect on the motion of the objects in the system.

4. The interaction of two forces can have a ripple effect on the motion of an object. If two forces act on an object in opposite directions, they will cancel each other out. If they act in the same

direction, they will add together. This interaction of forces can have a ripple effect on the object's velocity and momentum.

The ripple effect in electronics refers to the way in which one electrical event or change can have a series of consequences or impacts on other related events or phenomena. Here are a few examples of the ripple effect in electronics:

1. A change in the voltage of an electrical circuit can have a ripple effect on the current flowing through it. The voltage of an electrical circuit is the driving force that pushes the current through it. If you increase the voltage of a circuit, you will also increase the current flowing through it. This change in voltage can have a ripple effect on the power consumption of the circuit.

2. A change in the resistance of an electrical circuit can have a ripple effect on the current flowing through it. Resistance is the opposition to the flow of current in an electrical circuit. If you increase the resistance of a circuit, you will decrease the current flowing through it. This change in resistance can have a ripple effect on the power consumption of the circuit.

3. A change in the capacitance of an electrical circuit can have a ripple effect on the voltage across it. Capacitance is the ability of a circuit to store electric charge. If you increase the capacitance of a circuit, you will also increase the voltage across it. This change in capacitance can have a ripple effect on the power consumption of the circuit.

4. A change in the inductance of an electrical circuit can have a ripple effect on the current flowing through it. Inductance is the property of an electrical circuit that opposes a change in current. If you increase the inductance of a circuit, you will decrease the current flowing through it. This change in inductance can have a ripple effect on the power consumption of the circuit.

A chain reaction is a sequence of events in which the occurrence of one event causes another event to occur, which in turn causes another event to occur, and so on. This effect can be observed in various fields, including chemistry, physics, and biology.

One example of a chain reaction can be seen in nuclear fission. In nuclear fission, a neutron collides with a nucleus of a heavy atom, such as uranium or plutonium, which causes the nucleus to split into two smaller nuclei and release more neutrons. These neutrons can then collide with other nuclei and cause them to split as well, resulting in a chain reaction. This chain reaction is the basis of nuclear power and can also be used to make atomic bombs.

Another example of a chain reaction can be seen in chemistry. A chain reaction in chemistry is a sequence of chemical reactions in which the products of one reaction become reactants for the next reaction. A common example of a chain reaction in chemistry is combustion. In combustion, a fuel reacts with oxygen to produce heat, water vapor, and carbon dioxide. This reaction also produces heat which can cause the surrounding air molecules to heat up and increase the rate of the reaction.

Another example is a chain reaction in biology, where a small change in the population of one species can cause a chain of events that lead to a change in the population of other species. For example, a decrease in the population of predators, such as wolves, can lead to an increase in the population of their prey, such as deer. This in turn can lead to an increase in the population of plants that the deer feed on, which can then lead to a change in the population of insects and other animals that rely on those plants for food, and so on.

In summary, a chain reaction is a sequence of events in which one event causes another event, which in turn causes another event, and so on. It can be observed in various fields such as chemistry, physics, and biology, and it highlights the interconnectedness and interdependence of different systems.

The butterfly effect is a concept in chaos theory which states that a small change in initial conditions can lead to vastly different outcomes. The name of the effect is derived from the idea that the flapping of a butterfly's wings in Brazil could set off a chain of events that ultimately leads to a tornado in Texas.

An example of the butterfly effect can be seen in the weather. A small change in the temperature or wind direction in one location can set off a chain of events that leads to a completely different weather pattern in another location. For example, a storm system that develops in the Pacific Ocean can be affected by the temperature and wind patterns over Mexico, which in turn can be affected by the temperature and wind patterns over the Gulf of Mexico, and so on.

Another example of the butterfly effect can be seen in finance. A small change in the interest rate or exchange rate in one country can lead to a chain of events that results in a completely different financial outcome in another country. For example, a change in the interest rate in the United States can affect the value of the dollar, which in turn can affect the stock market in Japan, which can then affect the economy in Europe, and so on.

In summary, the butterfly effect is the phenomenon by which small causes can have large effects, particularly in complex systems. It highlights the sensitive dependence on initial conditions in nonlinear systems and the importance of considering all possible inputs and their interactions when making predictions or decisions.

The snowball effect, also known as the cumulative effect or the domino effect, refers to a process in which an initial event or action leads to a series of similar events or actions, each of which adds to the momentum of the process and makes it continue to grow. The term "snowball effect" comes from the idea that a small snowball, when rolled down a hill, will gather more snow as it goes, becoming larger and larger.

One example of the snowball effect is seen in personal debt. When a person takes on a small amount of debt, such as a credit card balance, it can be manageable to pay off. However, if that person continues to take on more debt, such as a car loan or a mortgage, it can become increasingly difficult to pay off all of the debt, as the minimum payments on each loan add up. As the debt grows, it can become harder to make the payments, which can lead to late fees and higher interest rates, further adding to the debt and making it even harder to pay off.

Another example of the snowball effect can be seen in business. A small company that starts by making a profit can use that profit to invest in new equipment, new products or services, and new employees. As the company grows, it can generate more revenue, which in turn can be used to invest in more growth, leading to an expanding customer base, more sales, and more profit. This process can continue to grow and create a virtuous cycle which can help the company to become more successful over time.

Another example of the snowball effect can be seen in public opinion. When a small number of people hold a certain viewpoint, it may not have a significant impact on the overall opinion of the population. However, as more and more people adopt that viewpoint, it can gain momentum and become more widespread, leading to a shift in the overall opinion of the population. This can be observed in the way news stories, political campaigns and social movements can spread and grow over time.

In summary, the snowball effect refers to a process in which an initial event or action leads to a series of similar events or actions, each of which adds to the momentum of the process and makes it continue to grow. It can be observed in various fields such as finance, business and public opinion and highlights the power of small events or actions to grow and become more significant over time.

The multiplier effect refers to the idea that an initial increase in spending can lead to a larger overall increase in economic activity. This occurs because the initial increase in spending causes a chain reaction of additional spending as the additional income generated by the initial spending is re-spent by recipients.

Here is an example of the multiplier effect at work:

1. A consumer spends $100 at a local store.
2. The store owner takes a portion of the $100 in profit and uses it to pay their employees.
3. The employees use their additional income to buy goods and services from other businesses.
4. The businesses that receive the additional spending use a portion of it to pay their employees, and so on.

As this chain reaction continues, the initial $100 of spending can lead to an overall increase in economic activity that is greater than $100. The size of the multiplier effect depends on a variety of factors, including the marginal propensity to consume (how much of an increase in income is spent rather than saved), the marginal propensity to import (how much of the increased spending goes towards imported goods rather than domestically produced goods), and the extent to which the increase in spending leads to increases in production (which can create additional jobs and income).

Another example of the multiplier effect is when the government increases spending on infrastructure projects. This can lead to an increase in economic activity as the government hires workers to complete the projects, and the workers use their additional income to buy goods and services from local businesses. The businesses, in turn, may use their additional revenue to hire more employees or expand their operations, further increasing economic activity.

The term "cascading effect" is often used to describe a situation where a small change in one area of a system can lead to larger and larger changes in other areas of the system. This can occur in a variety of natural systems, such as ecosystems, weather patterns, and geological processes.

Here are a few examples of the cascading effect in nature:

1. In an ecosystem, a small change in the population of one species can lead to changes in the populations of other species. For example, if the population of predators decreases, the population of their prey may increase, leading to a decrease in the populations of herbivores that depend on the prey as a food source. This chain reaction can continue, affecting the populations of other species in the ecosystem.

2. In weather patterns, a small change in temperature or atmospheric pressure can lead to larger and larger changes in the distribution and intensity of precipitation, wind, and other meteorological phenomena. For example, a small increase in sea surface temperature can lead to the formation of a tropical cyclone, which can have significant impacts on the surrounding area.

3. In geological processes, a small change in the stability of a slope or the level of ground water can lead to larger and larger changes in erosion,

landslides, and other geomorphic processes. For example, a small increase in the water level in a lake can cause erosion along the shoreline, which can lead to landslides on the surrounding hillsides.

The term "cascading effect" is often used to describe a situation where a small change in one area of a system can lead to larger and larger changes in other areas of the system. This can occur in a variety of physical systems, such as mechanical systems, electrical circuits, and chemical reactions.

Here are a few examples of the cascading effect in the physical sciences:

1. In mechanical systems, a small change in the position or velocity of one object can lead to larger and larger changes in the positions and velocities of other objects. For example, a small push on a domino can cause it to knock over several other dominos, each of which pushes several more dominos, and so on.

2. In electrical circuits, a small change in the current or voltage of one component can lead to larger and larger changes in the currents and voltages of other components. For example, a small increase in the voltage of a battery can cause a larger current to flow through a circuit, causing a larger voltage drop across a resistor, which can cause even more current to flow, and so on.

3. In chemical reactions, a small change in the concentration of one reactant can lead to larger and larger changes in the concentrations of other reactants and products. For example, a small increase in the concentration of a catalyst can cause a chemical reaction to proceed much

more quickly, leading to larger and larger changes in the concentrations of the reactants and products.

The term "cascading effect" can also be used to describe situations in social systems where a small change in one area can lead to larger and larger changes in other areas. This can occur in a variety of social systems, such as economic systems, political systems, and social networks.

Here are a few examples of the cascading effect in social science:

1. In economic systems, a small change in the price of a good or service can lead to larger and larger changes in the demand for the good or service, as well as the prices of related goods and services. For example, a small increase in the price of gasoline can lead to a decrease in the demand for gasoline, which can cause the price of gasoline to decrease even further. This chain reaction can continue, affecting the prices of other goods and services that are related to gasoline.

2. In political systems, a small change in the behavior or attitudes of one group can lead to larger and larger changes in the behavior or attitudes of other groups. For example, a small increase in the support for a particular political party can lead to an increase in the likelihood that other people will also support the party, which can cause even more people to support the party, and so on.

3. In social networks, a small change in the behavior or attitudes of one person can lead to

larger and larger changes in the behavior or attitudes of other people. For example, a small increase in the popularity of a particular product or service can cause more and more people to adopt the product or service, leading to even more popularity, and so on. This process is often referred to as "social contagion."

The term "cascading effect" can be used to describe situations in which a small change in one aspect of human nature leads to larger and larger changes in other aspects of human nature. Here are a few examples of the cascading effect in human nature:

1. A small change in a person's beliefs or attitudes can lead to larger and larger changes in their behavior. For example, a small change in a person's belief about the importance of exercising regularly can lead to a change in their behavior such as starting to exercise more frequently, which can lead to even more changes in their beliefs and attitudes, such as feeling healthier and more confident.

2. A small change in a person's social environment can lead to larger and larger changes in their social relationships and social roles. For example, a small change in a person's job or living situation can lead to a change in their social network, such as making new friends or joining new social groups, which can lead to even more changes in their social environment and relationships.

3. A small change in a person's physical environment can lead to larger and larger changes in their habits and routines. For example, a small change in a person's daily commute, such as taking a different route to work, can lead to a change in their schedule, such as leaving for work at a different time,

which can lead to even more changes in their habits and routines.

In astronomy, the term "cascading effect" can refer to a situation where a small change in one celestial body or system leads to larger and larger changes in other celestial bodies or systems. Here are a few examples of the cascading effect in astronomy:

1. A small change in the orbit of a planet or moon can lead to larger and larger changes in the orbits of other celestial bodies. For example, a small change in the orbit of a planet can cause it to pass closer to or farther from its star, which can lead to changes in the planet's climate and the likelihood of it being habitable. This, in turn, can affect the evolution of life on the planet and the likelihood of it being habitable by humans.

2. A small change in the mass of a celestial body can lead to larger and larger changes in its gravitational influence on other celestial bodies. For example, a small increase in the mass of a black hole can cause it to capture more matter, leading to an even larger increase in mass, which can cause it to capture even more matter, and so on.

3. A small change in the intensity of a star's radiation can lead to larger and larger changes in the temperature and composition of the star's outer layers. For example, a small increase in the temperature of a star's photosphere can cause it to become more luminous, leading to a larger increase in

temperature, which can cause it to become even more luminous, and so on. This process is known as a "runaway feedback loop."

In physics, the term "cascading effect" can refer to a situation where a small change in one physical system leads to larger and larger changes in other physical systems. Here are a few examples of the cascading effect in physics:

1. A small change in the energy of a system can lead to larger and larger changes in the system's temperature and the motion of its particles. For example, a small increase in the kinetic energy of a gas molecule can cause it to collide with other gas molecules more frequently, leading to an increase in the overall temperature of the gas and the average kinetic energy of its molecules.

2. A small change in the shape of a solid can lead to larger and larger changes in the solid's stress and strain. For example, a small increase in the pressure on a metal bar can cause it to deform, leading to an increase in the bar's internal stress and strain, which can cause it to deform even more, and so on.

3. A small change in the flow of a fluid can lead to larger and larger changes in the fluid's velocity and pressure. For example, a small increase in the velocity of a fluid in a narrow channel can cause an increase in the fluid's pressure, which can cause the velocity to increase even more, leading to an even larger increase in pressure, and so on. This process is known as a "hydraulic jump."

In physics, the term "cascading effect" can refer to a situation where a small change in one physical force leads to larger and larger changes in other physical forces. Here are a few examples of the cascading effect with physical forces:

1. A small change in the gravitational force acting on an object can lead to larger and larger changes in the object's velocity and position. For example, a small increase in the gravitational force on a falling object can cause it to accelerate faster, leading to a larger change in velocity, which can cause it to fall even faster, and so on.

2. A small change in the electrical force between charged particles can lead to larger and larger changes in the particles' charges and the strength of the electrical field. For example, a small increase in the electrical force between two charged particles can cause them to exchange more electrons, leading to a larger change in their charges, which can cause the electrical force between them to increase even more, and so on.

3. A small change in the magnetic force between magnetic poles can lead to larger and larger changes in the strength of the magnetic field and the motion of magnetic particles. For example, a small increase in the magnetic force between two magnetic poles can cause more

magnetic particles to be attracted to the poles, leading to a larger change in the strength of the magnetic field, which can cause even more magnetic particles to be attracted to the poles, and so on.

In electronics, the term "cascading effect" can refer to a situation where a small change in one circuit element leads to larger and larger changes in other circuit elements. Here are a few examples of the cascading effect in electronics:

1. A small change in the resistance of a resistor can lead to larger and larger changes in the current flowing through the resistor and the voltage across it. For example, a small increase in the resistance of a resistor in a circuit can cause a decrease in the current flowing through the resistor, leading to a decrease in the voltage across it, which can cause even more of a decrease in the current flowing through the resistor, and so on.

2. A small change in the capacitance of a capacitor can lead to larger and larger changes in the charge stored on the capacitor and the voltage across it. For example, a small increase in the capacitance of a capacitor in a circuit can cause an increase in the charge stored on the capacitor, leading to an increase in the voltage across it, which can cause even more of an increase in the charge stored on the capacitor, and so on.

3. A small change in the transconductance of a transistor can lead to larger and larger changes in the current flowing through the transistor and the voltage across it. For example, a small increase in the transconductance of a transistor

in a circuit can cause an increase in the current flowing through the transistor, leading to an increase in the voltage across it, which can cause even more of an increase in the current flowing through the transistor, and so on.

The butterfly effect is a concept in chaos theory that describes the idea that small, seemingly insignificant events can have large, unforeseen consequences. It is often used to illustrate the sensitivity of complex systems to initial conditions. The term is based on the metaphor of a butterfly flapping its wings in one part of the world causing a tornado to occur in another part of the world.

Here are a few examples of the butterfly effect in action:

1. A small change in the initial conditions of a weather system, such as the temperature or humidity of the air, can lead to large differences in the path and intensity of a storm. For example, a small change in the temperature of the air in one location can cause a change in the direction of the wind, which can cause a storm to develop in a different location or with a different intensity than it would have otherwise.

2. A small change in the initial conditions of a financial system, such as the value of a currency or the level of interest rates, can lead to large differences in the performance of financial markets. For example, a small change in the value of a currency can cause a change in the relative attractiveness of different investments, leading to changes in the demand for those investments and the prices at which they trade.

3. A small change in the initial conditions of a social system, such as the behavior of a single individual, can lead to large differences in the behavior of the entire system. For example, a small change in the behavior of a single person can cause a change in the behavior of those around them, which can cause a change in the behavior of even more people, and so on. This process is often referred to as "social contagion."

The natural state of a system refers to the state that a system tends to return to when it is not being subjected to external forces or influences. In other words, it is the state of balance or stability that a system tends to seek when it is not being disturbed.

Here are a few examples of the natural state of a system:

1. Mechanical equilibrium: In mechanics, the natural state of a system is one of mechanical equilibrium, where the forces acting on the system are balanced and the system is not accelerating. For example, a book sitting on a table is in its natural state when it is at rest and not being disturbed.

2. Thermal equilibrium: In thermodynamics, the natural state of a system is one of thermal equilibrium, where the energy absorbed by the system is equal to the energy it radiates into its surroundings. For example, a cup of hot coffee will eventually reach thermal equilibrium with the room it is in, at which point it will no longer be gaining or losing heat.

3. Chemical equilibrium: In chemistry, the natural state of a system is one of chemical equilibrium, where the rates of the forward and reverse reactions are equal and the concentrations of reactants and products are constant. For example, in the reaction between hydrogen gas and oxygen gas to form water, the system will eventually reach chemical equilibrium when the

rate at which hydrogen and oxygen molecules combine to form water is equal to the rate at which water molecules break down into hydrogen and oxygen.

Duality in general nature refers to the idea that many natural phenomena can be described and understood in multiple ways, and that these different perspectives are interconnected and equivalent. Here are a few examples of duality in nature:

Wave-particle duality: This concept states that particles, such as electrons, can exhibit both wave-like and particle-like behavior depending on how they are observed. For example, electrons can act as waves when passing through a double-slit experiment, but can act as particles when interacting with a detector.

Electromagnetic duality: This refers to the idea that electricity and magnetism are two different aspects of the same phenomenon. For example, a changing electric field generates a magnetic field, and a changing magnetic field generates an electric field. This is described by Maxwell's equations.

Duality in thermodynamics: The Carnot cycle, which is a theoretical model of an idealized heat engine, illustrates the relationship between work and heat. The heat absorbed during the isothermal expansion is equal to the work done during the adiabatic compression and vice versa.

Holographic principle: This principle states that the information content of a region of space is equivalent to the information content on its boundary. It implies that the physics in the bulk of a region is equivalent to the physics on its boundary.

Quantum field theory and general relativity: These two theories are the foundation of modern physics, but they describe different aspects of nature. While general relativity describes gravity, quantum field theory describes the other three fundamental forces of nature.

These examples demonstrate that duality in nature refers to the idea that many natural phenomena can be described and understood in multiple ways and that these different perspectives are interconnected and equivalent. Duality in nature is a fundamental concept in physics and helps to uncover the deeper connections between seemingly different phenomena.

Duality is a concept in mathematics and physics that refers to the relationship between two seemingly different systems or concepts that are actually equivalent or interchangeable. In other words, duality suggests that two seemingly distinct things are actually the same thing when viewed from a different perspective. Here are a few examples of duality:

In linear programming, the primal problem and the dual problem are two different formulations of the same optimization problem. The primal problem is the original problem being solved, while the dual problem is a reformulation of the primal problem that can be used to find alternative solutions.

In electrical engineering, the concept of circuit duality states that any electric circuit can be transformed into a dual circuit by interchanging the roles of voltage and current.

In geometry, the concept of a point-line duality states that any statement about points in a plane can be transformed into a statement about lines and vice versa.

In physics, there is a wave-particle duality which states that particles, such as electrons, can exhibit both wave-like and particle-like behavior depending on how they are observed.

In computer science, the concept of a Turing machine and a lambda calculus are dual to each other.

These are just a few examples of duality, but the concept can be applied in many different fields and disciplines. The key idea is that two seemingly different systems or concepts are actually equivalent or interchangeable.

Duality is a concept that is present in many different areas of physics, and it is particularly important in quantum mechanics. In quantum mechanics, duality refers to the idea that particles can exhibit both wave-like and particle-like behavior, depending on how they are observed. This is known as wave-particle duality.

One of the most famous examples of wave-particle duality is the double-slit experiment. In this experiment, a beam of electrons is shot at a screen with two slits in it. On the other side of the screen, a detector is placed to measure the electrons that pass through the slits. When the electrons are observed one at a time, they behave like particles and produce a pattern on the detector that is consistent with electrons passing through two distinct slits. However, when the electrons are not observed, they behave like waves, and produce an interference pattern on the detector that is consistent with waves passing through two slits.

Another example of wave-particle duality can be seen in the behavior of photons, which are the particles that make up light. Photons can exhibit both wave-like and particle-like behavior, depending on how they are observed. For example, when photons are sent through a double-slit experiment, they produce an interference pattern on a detector, which is consistent with their wave-like behavior. However, when photons are detected one at a time, they produce a pattern on the detector that is consistent with their particle-like behavior.

Wave-particle duality is not limited to electrons and photons, it also applies to other particles, such as protons, neutrons, and even atoms and molecules.

In addition to wave-particle duality, there are other types of dualities in nature. For example, in particle physics, there is a strong-weak duality, which states that the strong nuclear force (the force that holds protons and neutrons together in the nucleus) and the weak nuclear force (the force responsible for certain types of radioactive decay) are different manifestations of the same underlying force.

In string theory, there is a duality between a theory in which particles are point-like and a theory in which particles are one-dimensional extended objects (strings).

In summary, duality is a concept that is present in many areas of physics, and it refers to the idea that particles can exhibit both wave-like and particle-like behavior depending on how they are observed. Additionally, there are other types of dualities that exist in nature, such as strong-weak duality and string theory dualities.

Duality in the economy refers to the idea that economic systems can be described and understood from multiple perspectives, and that these different perspectives are interconnected and equivalent. Here are a few examples of duality in the economy:

Labor market duality: This concept states that there are two types of labor markets: formal and informal. Formal labor markets are characterized by permanent, salaried employment, while informal labor markets are characterized by temporary, unsalaried, and unregulated employment.

Product market duality: This concept states that there are two types of product markets: formal and informal. Formal product markets are characterized by regulated and standardized products, while informal product markets are characterized by unregulated and diverse products.

Duality in monetary policy: Central banks use monetary policy to control the money supply and interest rates in an economy. However, monetary policy can also be used to control exchange rates, inflation and interest rates by the use of open market operations, discount rate, and reserve requirements.

Duality in financial markets: Financial markets can be divided into primary markets, where new securities are issued, and secondary markets, where securities are traded after they have been issued. Primary markets are characterized by equity issuance, while secondary markets are characterized by bond trading.

Duality in economic growth: Economic growth can be measured by either GDP or by the Human Development Index (HDI). GDP measures the monetary value of goods and services produced in a country, while HDI measures the well-being of the population by taking into account factors such as health, education, and standard of living.

These examples demonstrate that duality in the economy refers to the idea that economic systems can be described and understood from multiple perspectives, and that these different perspectives are interconnected and equivalent. Duality in the economy is a fundamental concept that helps to understand the complex interactions between different economic actors and institutions.

Duality in physical science refers to the idea that many physical phenomena can be described and understood in multiple ways, and that these different perspectives are interconnected and equivalent. Here are a few examples of duality in physical science:

Wave-particle duality: This concept states that particles, such as electrons, can exhibit both wave-like and particle-like behavior depending on how they are observed. For example, electrons can act as waves when passing through a double-slit experiment, but can act as particles when interacting with a detector.

Electromagnetic duality: This refers to the idea that electricity and magnetism are two different aspects of the same phenomenon. For example, a changing electric field generates a magnetic field, and a changing magnetic field generates an electric field. This is described by Maxwell's equations.

Thermal-mechanical duality: This concept refers to the idea that the laws of thermodynamics, which govern the behavior of heat and energy, can be mapped onto the laws of mechanics, which govern the behavior of matter and motion.

Duality in Quantum Mechanics: Quantum Mechanics is a branch of physics that provides a mathematical framework to describe the behavior of particles at the atomic and subatomic level. Heisenberg Uncertainty Principle, which states that the more precisely the position of a particle is known, the less precisely its momentum can be known and vice versa, is an example of duality in Quantum Mechanics.

String theory and M-theory: String theory and M-theory are two different theories that attempt to unify the fundamental forces of nature, including gravity, electromagnetism, and the strong and weak nuclear forces. They propose that the universe is made up of tiny, one-dimensional "strings" or "membranes" that vibrate at different frequencies and interact with each other to create the universe we see.

These examples demonstrate that duality in physical science refers to the idea that many physical phenomena can be described and understood in multiple ways, and that these different perspectives are interconnected and equivalent. Duality in physical science is a fundamental concept that helps to understand the complex interactions between different physical phenomena and to uncover the deeper connections between seemingly different phenomena.

Duality in social science refers to the idea that two seemingly opposite concepts or phenomena are actually closely related or interconnected. One example of duality in social science is the relationship between structure and agency. Structure refers to the larger societal and cultural forces that shape our lives and limit our choices, while agency refers to the individual's ability to make choices and act on their own behalf. These two concepts are often seen as opposing forces, but in reality, they are interconnected and mutually dependent. For example, an individual's agency is shaped by the structure of the society they live in, and their choices and actions can also shape the structure of that society.

Another example of duality in social science is the relationship between power and resistance. Power refers to the ability of individuals or groups to control or influence others, while resistance refers to the actions taken by individuals or groups to challenge or push back against that power. These two concepts are often seen as opposing forces, but in reality, they are interconnected. For example, power is often maintained and reinforced through resistance, and resistance is often necessary for individuals or groups to gain power.

Yet another example is the relationship between culture and identity. Culture refers to the shared beliefs, values, customs, behaviors, and artifacts that characterize a group or society, while identity refers to the individual's sense of self and how they see themselves in relation to the culture they are part of. These two concepts are often seen as separate, but in reality, they are interconnected. For example, an individual's identity is shaped by the culture they are part of, and their identity can also shape the culture they are part of.

In summary, duality in social science refers to the idea that two seemingly opposite concepts or phenomena are actually closely related or interconnected. Examples of duality include the relationship between structure and agency, power and resistance, and culture and identity.

Duality in human nature refers to the idea that individuals possess both positive and negative traits, or that there are opposing forces within human nature that can manifest in different ways. This can be viewed as the coexistence of good and evil, or the balance between self-interest and compassion.

One example of duality in human nature is the relationship between self-interest and altruism. Self-interest refers to the idea that individuals act in a way that benefits themselves, while altruism refers to the idea that individuals act in a way that benefits others. These two concepts are often seen as opposing forces, but in reality, they can coexist in the same individual. For example, an individual may act in a self-interested manner in some situations, such as pursuing their own career goals, but also act in an altruistic manner in other situations, such as volunteering in their community.

Another example of duality in human nature is the relationship between aggression and cooperation. Aggression refers to behavior that is intended to harm others, while cooperation refers to behavior that is intended to work together with others. These two concepts are often seen as opposing forces, but in reality, they can coexist in the same individual. For example, an individual may act aggressively in some situations, such as competing in a sport, but also act cooperatively in other situations, such as working on a team project.

Yet another example is the relationship between rationality and emotionality. Rationality refers to the ability to think logically, while emotionality refers to the ability to feel and express emotions. These two concepts are often seen as separate, but in reality, they are interconnected and have an impact on each other. For example, an individual's rationality can be influenced by their emotions, and their emotions can be influenced by their rationality.

In summary, duality in human nature refers to the idea that individuals possess both positive and negative traits, or that there are opposing forces within human nature that can manifest in different ways. Examples of duality include the relationship between self-interest and altruism, aggression and cooperation, and rationality and emotionality.

Duality in astronomy refers to the idea that there are two seemingly opposing or distinct phenomena that are actually closely related or interconnected. One example of duality in astronomy is the relationship between black holes and neutron stars. Black holes are extremely dense regions of space with extremely strong gravitational pull, from which not even light can escape. Neutron stars, on the other hand, are extremely dense, compact stars that are formed from the collapsed cores of massive stars. While black holes and neutron stars may seem vastly different, they are both formed from the collapse of massive stars and are therefore closely related.

Another example of duality in astronomy is the relationship between dark matter and dark energy. Dark matter is a hypothetical form of matter that is thought to make up approximately 85% of the matter in the universe, while dark energy is a hypothetical form of energy that is thought to make up approximately 68% of the universe. These two concepts may seem vastly different, but they are both thought to play a crucial role in the formation and expansion of the universe.

A third example of duality in astronomy is the relationship between supernovae and gamma-ray bursts. Supernovae are the explosive death of a star, while gamma-ray bursts are intense and short bursts of gamma-ray radiation. Both supernovae and gamma-ray bursts are caused by the collapse and explosion of massive stars, and both are among the most energetic and luminous events in the universe.

In summary, duality in astronomy refers to the idea that there are two seemingly opposing or distinct phenomena that are actually closely related or interconnected. Examples of duality in astronomy include the relationship between black holes and neutron stars, dark matter and dark energy, and supernovae and gamma-ray bursts.

Duality in physics refers to the idea that certain physical phenomena can be described by two seemingly different but equivalent models or frameworks. One of the most well-known examples of duality in physics is wave-particle duality, which states that certain physical entities, such as electrons, can exhibit both wave-like and particle-like behavior. This duality is described by the wave-particle duality theory, which shows that the behavior of subatomic particles can be described by the principles of both quantum mechanics and wave mechanics.

Another example of duality in physics is the relationship between general relativity and quantum mechanics. General relativity is a theory of gravity that describes the behavior of large-scale physical systems, while quantum mechanics is a theory of physics that describes the behavior of subatomic particles. These two theories seem vastly different, but they are both needed to explain the behavior of physical systems in the universe.

A third example of duality in physics is the relationship between electric and magnetic fields. Electric fields and magnetic fields are two different types of fields that are related to each other through the principle of electromagnetic duality, which states that the electric and magnetic fields are interchangeable. This duality is described by the Maxwell's equations, which unify the laws of electricity and magnetism.

Another example is the T-duality which is a symmetry of string theory, it states that the physics of a string theory is identical when the radius of the compactified dimension is changed to its reciprocal.

In summary, duality in physics refers to the idea that certain physical phenomena can be described by two seemingly different but equivalent models or frameworks. Examples of duality in physics include wave-particle duality, the relationship between general relativity and quantum mechanics, the relationship between electric and magnetic fields and T-duality in string theory.

Duality with physical forces refers to the idea that certain physical forces can be described by two seemingly different but equivalent models or frameworks. One example of this is electromagnetic duality, which states that the electric and magnetic fields are interchangeable. This duality is described by the Maxwell's equations, which unify the laws of electricity and magnetism, and can be observed in phenomena such as electromagnetic waves.

Another example is the strong-weak duality, which states that the strong nuclear force and the weak nuclear force, the two fundamental forces that govern the behavior of subatomic particles, can be described by a single unified theory. This duality is described by the electroweak theory, which unifies the weak nuclear force and the electromagnetic force.

A third example is the duality between gravity and thermodynamics. The principle of equivalence states that the gravitational force and the acceleration are equivalent, and the laws of thermodynamics can be used to describe the properties of black holes. This duality is described by the equivalence principle and the laws of black hole thermodynamics.

In summary, duality with physical forces refers to the idea that certain physical forces can be described by two seemingly different but equivalent models or frameworks. Examples of duality with physical forces include electromagnetic duality, the strong-weak duality, and the duality between gravity and thermodynamics.

Duality in electronics refers to the idea that certain electronic devices or systems can be described by two seemingly different but equivalent models or frameworks. One example of duality in electronics is the relationship between analog and digital circuits. Analog circuits are electronic systems that operate on continuous signals, while digital circuits are electronic systems that operate on discrete signals. These two types of circuits may seem vastly different, but they are both used in electronic devices, and they are closely related. For example, an analog signal can be converted into a digital signal, and a digital signal can be converted into an analog signal through the use of analog-to-digital and digital-to-analog converters.

Another example of duality in electronics is the relationship between field-effect transistors (FETs) and bipolar junction transistors (BJTs). FETs are electronic devices that use a voltage applied to a gate electrode to control the current flowing through the device, while BJTs are electronic devices that use a current applied to a base electrode to control the current flowing through the device. These two types of transistors may seem vastly different, but they are both widely used in electronic devices and they have similar characteristics.

A third example of duality in electronics is the relationship between capacitors and inductors. Capacitors are electronic components that store electrical energy in an electric field, while inductors are electronic components that store electrical energy in a magnetic field. These two types of components may seem vastly different, but they are both widely used in electronic devices and they have similar characteristics.

In summary, duality in electronics refers to the idea that certain electronic devices or systems can be described by two seemingly different but equivalent models or frameworks. Examples of duality in electronics include the relationship between analog and digital circuits, field-effect transistors and bipolar junction transistors, and capacitors and inductors.

A cycle refers to a repeating pattern of events or processes. Cycles can be observed in many natural and man-made systems, and they often involve changes in the state of a system over time.

One example of a cycle can be seen in the seasons. The Earth's orbit around the sun and its tilt on its axis cause the climate to change over the course of a year, leading to the familiar cycles of spring, summer, fall, and winter. The cycles of the seasons are caused by the changing of the earth's distance from the sun and the tilt of the earth's axis which leads to changes in the amount of sunlight that different parts of the earth receive.

Another example of a cycle can be seen in the water cycle. The water cycle is the process by which water evaporates from the surface of the earth, forms clouds, and falls back to the surface as precipitation. The precipitation can be in the form of rain, snow, or hail. The water that falls to the surface can be stored in rivers, lakes, or underground aquifers and can eventually evaporate again, starting the cycle over.

Another example of a cycle can be seen in the business cycle. Business cycles refer to the natural fluctuations of an economy over time, characterized by periods of growth (expansion) and decline (recession). Business cycles are typically measured by changes in GDP, employment, and investment. The causes of these cycles can be a combination of factors such as changes in consumer demand, technological innovations, monetary policy and global economic conditions.

In summary, a cycle refers to a repeating pattern of events or processes. Cycles can be observed in many natural and man-made systems, including the seasons, the water cycle and the business cycle. They are caused by a variety of factors and often involve changes in the state of a system over time. Understanding cycles can help to predict and plan for future changes in a system, and can be useful for decision making in various fields.

A pendulum is a simple mechanical device consisting of a weight suspended from a fixed point that swings back and forth under the influence of gravity. The pendulum has been used in many applications, such as timekeeping and scientific experimentation.

The motion of a pendulum is a type of harmonic motion, meaning that it oscillates back and forth at a constant frequency, called the pendulum's natural frequency. The natural frequency of a pendulum is determined by its length, the gravitational field, and the mass of the bob (the weight at the end of the pendulum). The longer the pendulum, the slower its natural frequency.

One example of a pendulum is a grandfather clock. A grandfather clock uses a pendulum to keep time. The pendulum swings back and forth at a constant rate, which is regulated by an escapement mechanism. The escapement mechanism releases the pendulum at regular intervals, allowing it to swing back and forth and keep time.

Another example of a pendulum is a swing. A swing is a pendulum that is powered by the energy of the person sitting on it. The person sitting on the swing pushes it back and forth, giving it energy. The swing then oscillates back and forth at a natural frequency determined by the length of the ropes and the weight of the person sitting on it.

A third example of a pendulum is a seismometer. A seismometer is a device that measures ground vibrations, such as those caused by earthquakes. The seismometer uses a pendulum to measure the ground vibrations. The pendulum oscillates back and forth in response to the ground vibrations, and the motion of the pendulum is used to measure the amplitude and frequency of the vibrations.

In summary, a pendulum is a simple mechanical device consisting of a weight suspended from a fixed point that swings back and forth under the influence of gravity. The motion of a pendulum is a type of harmonic motion, meaning that it oscillates back and forth at a constant frequency. The natural frequency of a pendulum is determined by its length, the gravitational field, and the mass of the bob. Pendulum is used in various applications such as timekeeping, swings and seismometer.

Level theory is a psychological theory that proposes that individuals have different levels of needs that must be met in order for them to feel fulfilled and motivated. The theory is based on the idea that lower-level needs must be met before higher-level needs can be addressed.

The theory was first proposed by Abraham Maslow in his hierarchy of needs, which is often represented as a pyramid with the basic physiological needs at the bottom, followed by safety needs, love and belonging needs, esteem needs, and self-actualization needs at the top. Maslow proposed that individuals must satisfy their basic physiological and safety needs before they can focus on satisfying their social and esteem needs, and ultimately self-actualization.

An example of level theory can be seen in the workplace. An employee who is not being paid enough to meet their basic physiological needs, such as food and shelter, may be preoccupied with finding a new job or obtaining a raise, and may not be able to focus on higher-level needs such as career advancement or job satisfaction. Once their basic needs are met, the employee may be able to focus on the higher-level needs.

Another example of level theory can be seen in the context of education. A student who is struggling to meet their basic physiological needs, such as hunger or lack of sleep, may not be able to focus on their studies and may struggle to achieve their academic goals. Once their basic needs are met, they will be better able to focus on their education and achieve their academic goals.

In summary, Level theory is a psychological theory that proposes that individuals have different levels of needs that must be met in order for them to feel fulfilled and motivated. The theory is based on the idea that lower-level needs must be met before higher-level needs can be addressed. The theory has been applied in various fields such as workplace, education and personal development. Understanding the level theory could help individuals and organizations to better understand the needs of the people they work with, and to create an environment where individuals can be motivated and fulfilled.

In physics, the term "level" refers to a specific energy state that a system, such as an atom or molecule, can occupy. Levels are usually quantized, meaning that the energy can only take on specific discrete values, rather than any value within a range.

One example of levels in physics is the energy levels of an atom. Atoms have different energy levels, or orbitals, that the electrons can occupy. Each energy level is associated with a specific amount of energy, and an electron can only occupy a specific energy level if it has the corresponding amount of energy. When an electron jumps from one energy level to another, it either absorbs or emits a photon of energy.

Another example of levels in physics is the energy levels of a molecule. Molecules also have energy levels that the electrons can occupy. These energy levels are called rotational, vibrational and electronic levels. The rotational levels are associated with the rotation of the molecule, the vibrational levels with the vibrations of the atoms in the molecule, and the electronic levels with the electronic configuration of the molecule.

A third example of levels in physics is the energy levels of a quantum mechanical system, such as a particle in a box or a harmonic oscillator. These systems have a discrete set of energy levels that the particle can occupy. The energy levels depend on the properties of the system, such as the size of the box or the strength of the potential, and the particle can only occupy specific energy levels if it has the corresponding amount of energy.

In summary, In physics, the term "level" refers to a specific energy state that a system, such as an atom, molecule or a quantum mechanical system, can occupy. Levels are usually quantized, meaning that the energy can only take on specific discrete values, rather than any value within a range. The levels in physics can be observed in various systems such as atoms, molecules and quantum mechanical systems. Understanding the levels in a system can help to predict and explain the behavior of that system, and can be useful for decision making in various fields such as physics, chemistry and material science.

The law of diminishing returns, also known as the law of diminishing marginal returns, states that as the quantity of one input is increased while the quantities of all other inputs are held constant, a point will eventually be reached at which the marginal return from an additional unit of input will begin to decrease.

One example of the law of diminishing returns can be seen in agriculture. As a farmer increases the amount of fertilizer applied to a field, the yield of crops will initially increase. However, at some point, adding more fertilizer will no longer result in a significant increase in crop yield and may even decrease the yield if the fertilizer is applied in excess.

Another example of the law of diminishing returns can be seen in the manufacturing industry. As a factory increases the number of machines and workers, the output of the factory will initially increase. However, at some point, adding more machines and workers will no longer result in a significant increase in output and may even decrease the output if the machines and workers are not used efficiently.

A third example of the law of diminishing returns can be seen in the service industry. As a restaurant increases the number of servers, the number of customers served will initially increase. However, at some point, adding more servers will no longer result in a significant increase in the number of customers served and may even decrease the number of customers served if the servers are not utilized efficiently.

In summary, the law of diminishing returns states that as the quantity of one input is increased while the quantities of all other inputs are held constant, a point will eventually be reached at which the marginal return from an additional unit of input will begin to decrease. It can be observed in various fields such as agriculture, manufacturing and service industry. Understanding the law of diminishing returns can help organizations and individuals to make more efficient use of their resources and to optimize their production or service processes.

In mathematics, a maximum or maxima refers to the highest value that a function can achieve within a given domain. The point at which a function reaches its maximum value is known as a relative maximum or local maximum. A global maximum is the highest point of a function over the entire domain of the function. A function can have multiple local maxima but only one global maximum.

One example of a maximum can be seen in the study of temperature. The maximum temperature that a substance can reach is known as its boiling point. For example, the boiling point of water is 100 degrees Celsius (212 degrees Fahrenheit) at standard pressure.

Another example of a maximum can be seen in the study of population growth. The maximum population that an ecosystem can support is known as its carrying capacity. For example, a certain area of land may only be able to support a maximum of 100 deer, beyond which the population will start to decline due to a lack of resources.

A third example of a maximum can be seen in the study of optimization problems. In optimization problems, the goal is to find the maximum or minimum value of a function subject to certain constraints. For example, an airline may want to find the flight schedule that maximizes the number of passengers it can carry while minimizing the number of flights it needs to operate.

In summary, In mathematics, a maximum or maxima refers to the highest value that a function can achieve within a given domain. The point at which a function reaches its maximum value is known as a relative maximum or local maximum. A global maximum is the highest point of a function over the entire domain of the function. Maxima can be observed in various fields such as temperature, population growth and optimization problems. Understanding maxima can help individuals and organizations to make more efficient use of their resources and to optimize their processes.

In mathematics, a minimum or minima refers to the lowest value that a function can achieve within a given domain. The point at which a function reaches its minimum value is known as a relative minimum or local minimum. A global minimum is the lowest point of a function over the entire domain of the function. A function can have multiple local minima but only one global minimum.

One example of a minimum can be seen in the study of temperature. The minimum temperature that a substance can reach is known as its freezing point. For example, the freezing point of water is 0 degrees Celsius (32 degrees Fahrenheit) at standard pressure.

Another example of a minimum can be seen in the study of cost. The minimum cost that a company can incur to produce a certain product or service is known as its minimum efficient scale. For example, a certain production process may have a minimum cost of $1000 when producing 1000 units, but the cost per unit drops to $500 when producing 10000 units.

A third example of a minimum can be seen in the study of optimization problems. In optimization problems, the goal is to find the minimum or maximum value of a function subject to certain constraints. For example, an engineer may want to design a structure that minimizes the amount of materials used while maximizing its strength.

In summary, In mathematics, a minimum or minima refers to the lowest value that a function can achieve within a given domain. The point at which a function reaches its minimum value is known as a relative minimum or local minimum. A global minimum is the lowest point of a function over the entire domain of the function. Minima can be observed in various fields such as temperature, cost, and optimization problems. Understanding minima can help individuals and organizations to make more efficient use of their resources and to optimize their processes.

The phrase "feast or famine" is a colloquial expression that describes a situation in which there is a significant alternation between periods of abundance and scarcity. It is often used to describe a pattern of events or circumstances in which there are extreme fluctuations between having too much or too little of something.

One example of feast or famine can be seen in the stock market. A stock market that experiences a bull run, where stock prices are rising rapidly, may be considered to be in a state of "feast," while a bear market, where stock prices are falling, may be considered to be in a state of "famine."

Another example of feast or famine can be seen in the agricultural industry. A farmer may have a good harvest one year, resulting in an abundance of crops and a high income, followed by a poor harvest the next year, resulting in a scarcity of crops and a low income.

A third example of feast or famine can be seen in the business cycle. A business may have a period of high sales and high profits, followed by a period of low sales and low profits.

In summary, the phrase "feast or famine" is a colloquial expression that describes a situation in which there is a significant alternation between periods of abundance and scarcity. It can be observed in various fields such as stock market, agricultural industry and business cycle. Understanding the feast or famine cycle can help individuals and organizations to make better decisions, plan for the future and make more efficient use of their resources.

In physics, chemistry and economics, the concept of upper and lower limits refers to the maximum and minimum values that a physical or economic system can reach. Equilibrium refers to a state in which the forces acting on a system are in balance and the system is stable. Upper and lower limits with equilibrium can be observed in various systems.

One example of upper and lower limits with equilibrium can be seen in the study of chemical reactions. The reactants are the starting materials of a chemical reaction and the products are the resulting materials. The equilibrium state is the point at which the rate of the forward reaction (reactants to products) is equal to the rate of the reverse reaction (products to reactants) and the concentration of reactants and products remain constant. The upper limit is the maximum concentration of the products that can be reached and the lower limit is the minimum concentration of the reactants that can be reached.

Another example of upper and lower limits with equilibrium can be seen in the study of population growth. The carrying capacity is the upper limit of the population that an ecosystem can support. The population will grow until it reaches the carrying capacity and then will stabilize at equilibrium. The lower limit is the minimum population that the ecosystem can support without going extinct.

A third example of upper and lower limits with equilibrium can be seen in the study of the market economy. The market price is the point at which the quantity supplied of a good equals the quantity demanded. The upper limit is the maximum price that buyers are willing to pay and the lower limit is the minimum price that sellers are willing to accept.

In summary, In physics, chemistry and economics, the concept of upper and lower limits refers to the maximum and minimum values that a physical or economic system can reach. Equilibrium refers to a state in which the forces acting on a system are in balance and the system is stable. Upper and lower limits with equilibrium can be observed in various systems such as chemical reactions, population growth and market economy. Understanding the upper and lower limits with equilibrium can help individuals and organizations to make better decisions, plan for the future and make more efficient use of their resources.

The statement "the more a system approaches equilibrium, the more unstable it can become" is based on the idea that as a system approaches equilibrium, the forces that are maintaining that equilibrium become weaker, which can make the system more prone to disruptions or changes.

One example of this can be seen in the study of chemical reactions. As a chemical reaction approaches equilibrium, the difference in concentrations between the reactants and products becomes smaller, which means that the rate of the forward and reverse reactions becomes closer to each other. This can make the reaction more susceptible to small changes in the concentrations of reactants and products, which can cause the reaction to shift away from equilibrium.

Another example can be seen in the study of ecosystems. As an ecosystem approaches carrying capacity, the competition for resources becomes stronger and the population becomes more sensitive to disturbances such as disease or changes in the environment. This can cause the population to fluctuate more and make the ecosystem more unstable.

A third example can be seen in the study of the economy. As an economy approaches market equilibrium, the difference in prices and quantities between supply and demand becomes smaller, which means that small changes in demand or supply can cause the market to shift away from equilibrium. This can make the economy more susceptible to fluctuations and make it more unstable.

In summary, the statement "the more a system approaches equilibrium, the more unstable it can become" is based on the idea that as a system approaches equilibrium, the forces that are maintaining that equilibrium become weaker, which can make the system more prone to disruptions or changes. This can be observed in various systems such as chemical reactions, ecosystems, and economy. Understanding this concept can help individuals and organizations to anticipate and prepare for potential disruptions, and make better decisions.

The statement "natural variation keeps stability" refers to the idea that in a dynamic system, small fluctuations or variations can help maintain stability and prevent large disruptions. These small variations can act as a buffer against external disturbances and prevent the system from reaching critical thresholds that would cause a collapse.

One example of natural variation keeping stability can be seen in the study of ecosystems. In a healthy ecosystem, there is a natural variation in the populations of different species. This variation can act as a buffer against disturbances such as disease or changes in the environment, preventing any one species from becoming too dominant and causing a collapse of the ecosystem. For instance, in the savannah grassland, the grazing pressure of large herbivores like elephants and zebras are kept in check by predation pressure of lions and hyenas.

Another example can be seen in the study of the economy. In a healthy economy, there is a natural variation in the prices and quantities of goods and services. This variation can act as a buffer against external disturbances such as changes in government policies or natural disasters, preventing any one market from becoming too unstable and causing a collapse of the economy.

A third example can be seen in the study of the stock market. In a healthy stock market, there is a natural variation in the prices of stocks. This variation can act as a buffer against external disturbances such as changes in economic conditions or geopolitical events, preventing any one stock from becoming too volatile and causing a collapse of the market.

In summary, the statement "natural variation keeps stability" refers to the idea that in a dynamic system, small fluctuations or variations can help maintain stability and prevent large disruptions. These small variations can act as a buffer against external disturbances and prevent the system from reaching critical thresholds that would cause a collapse. This concept can be observed in various systems such as ecosystems, economy and stock market. Understanding natural variation can help individuals and organizations to anticipate and prepare for potential disruptions, and make better decisions.

The bullwhip effect is a phenomenon that occurs in supply chain management when small changes in demand at the retail level lead to larger and larger changes in demand at the wholesale, distributor, and manufacturer levels. The bullwhip effect is caused by a combination of factors, including order batching, price fluctuations, and a lack of communication and coordination among supply chain partners.

One example of the bullwhip effect can be seen in the context of a retail store. When a customer buys a product, the store may place an order for more of that product from their supplier. However, if the store orders more of the product than is needed to meet the actual demand, it can lead to an oversupply and a buildup of inventory. This oversupply can then lead to markdowns and clearance sales, which can further disrupt the supply chain.

Another example of the bullwhip effect can be seen in the context of a manufacturing company. When a customer orders a product, the company may ramp up production to meet that demand. However, if the company produces more of the product than is needed to meet the actual demand, it can lead to an oversupply and a buildup of inventory. This oversupply can then lead to markdowns and clearance sales, which can further disrupt the supply chain.

A third example of the bullwhip effect can be seen in the context of a supply chain. When a customer orders a product, the supplier may order raw materials from their own supplier to manufacture the product. However, if the supplier orders more raw materials than is needed to meet the actual demand, it can lead to an oversupply and a buildup of inventory. This oversupply can then lead to markdowns and clearance sales, which can further disrupt the supply chain.

In summary, the bullwhip effect is a phenomenon that occurs in supply chain management when small changes in demand at the retail level lead to larger and larger changes in demand at the wholesale, distributor, and manufacturer levels. This can lead to an oversupply, a buildup of inventory, and a disruption of the supply chain. The bullwhip effect is caused by a combination of factors, including order batching, price fluctuations, and a lack of communication and coordination among supply chain partners. Understanding this concept can help individuals and organizations to anticipate and prepare for potential disruptions, and make better decisions.

Force field analysis is a tool that is used to evaluate a situation and identify the forces that are driving a change and the forces that are resisting it. The tool was developed by Kurt Lewin, a social psychologist, in the 1940s. The analysis is used to evaluate the potential for change in a system and to identify the factors that are likely to influence the success of that change.

The Force field analysis process is divided into two stages:

Identify the current situation and the forces driving the change and the forces resisting it.

Assess the relative strength of the driving and resisting forces, and identify the key factors that will influence the outcome of the change.

One example of Lewin's force field analysis can be seen in a business context, where a company wants to implement a new technology or process. The driving forces could include the potential benefits of the new technology, such as increased efficiency or cost savings, while the resisting forces could include the cost of implementing the new technology, and the potential disruption to existing processes. By identifying these forces, the company can evaluate the potential for change and develop a plan to address the key factors that will influence the success of the change.

Another example of Lewin's force field analysis can be seen in a political context, where a government wants to implement a new policy. The driving forces could include the potential benefits of the new policy, such as increased social welfare, while the resisting forces could include the cost of implementing the new policy, and the potential for opposition from interest groups. By identifying these forces, the government can evaluate the potential for change and develop a plan to address the key factors that will influence the success of the policy.

In summary, Force field analysis is a tool that is used to evaluate a situation and identify the forces that are driving a change and the forces that are resisting it. The tool was developed by Kurt Lewin, a social psychologist, in the 1940s. Force field analysis helps to evaluate the potential for change in a system and to identify the factors that are likely to influence the success of that change. This can be applied in various fields such as business, political and social context to evaluate the potential for change and develop a plan accordingly.

The path of least resistance is a concept that refers to the easiest or most natural way for something to move or occur. It is often used to describe the behavior of a fluid, such as water, as it flows through a channel or pipe, or the behavior of an object as it moves through a field of forces. The path of least resistance is the path that requires the least amount of energy to traverse.

One example of the path of least resistance can be seen in the movement of water. When water flows through a channel, it will naturally follow the path that has the least resistance, such as the path with the least amount of obstacles or the path with the least amount of friction. This means that the water will flow in the direction that requires the least amount of energy to move.

Another example of the path of least resistance can be seen in the behavior of a charged particle in an electric field. A charged particle will naturally move in the direction of the electric field that requires the least amount of energy to move.

A third example of the path of least resistance can be seen in the behavior of people. People tend to follow the path that requires the least amount of effort or resistance, whether it's in their personal or professional lives. For example, a person might choose to take the easiest route to work, even if it takes longer, because it requires less effort than a more difficult route.

In summary, the path of least resistance is a concept that refers to the easiest or most natural way for something to move or occur. It is often used to describe the behavior of a fluid, such as water, as it flows through a channel or pipe, or the behavior of an object as it moves through a field of forces. The path of least resistance is the path that requires the least amount of energy to traverse. This concept can be observed in various fields such as water flow, electric field and human behavior. Understanding the path of least resistance can help individuals and organizations to make better decisions, plan for the future and make more efficient use of their resources.

The hump of resistance is a term used to describe the point at which an individual or organization faces the greatest resistance when trying to implement a change. It is a concept that refers to the point in the change process where the most obstacles or resistance are encountered.

One example of the hump of resistance can be seen in the context of organizational change. A company may want to implement a new process or technology, but the employees may resist the change due to factors such as fear of the unknown or a lack of understanding of the benefits of the change. The hump of resistance occurs when the company is trying to overcome this resistance and implement the change.

Another example of the hump of resistance can be seen in the context of personal change. An individual may want to make a change in their life, such as quitting smoking or losing weight. The hump of resistance occurs when the individual is trying to overcome the obstacles and temptations that are preventing them from making the change.

A third example of the hump of resistance can be seen in the context of a political change. A government may want to implement a new policy, but the citizens may resist the change due to factors such as lack of trust in the government or lack of understanding of the benefits of the change. The hump of resistance occurs when the government is trying to overcome this resistance and implement the change.

In summary, The hump of resistance is a term used to describe the point at which an individual or organization faces the greatest resistance when trying to implement a change. It is a concept that refers to the point in the change process where the most obstacles or resistance are encountered. This concept can be observed in various fields such as organizational change, personal change and political change. Understanding the hump of resistance can help individuals and organizations to anticipate and prepare for potential obstacles and resistance, and make better decisions.

The law of attraction is a popular concept in self-help and personal development circles that suggests that we can attract positive experiences and outcomes into our lives by focusing our thoughts and feelings on them. The idea is that our thoughts and feelings create a kind of energy that attracts similar experiences and outcomes into our lives.

One example of the law of attraction can be seen in the context of career success. If an individual believes that they can be successful in their career and focuses their thoughts and feelings on that belief, they may be more likely to attract opportunities and experiences that lead to career success.

Another example of the law of attraction can be seen in the context of relationships. If an individual believes that they can have a fulfilling and loving relationship, and focuses their thoughts and feelings on that belief, they may be more likely to attract a compatible partner and build a healthy relationship.

A third example of the law of attraction can be seen in the context of financial success. If an individual believes that they can be financially successful and focuses their thoughts and feelings on that belief, they may be more likely to attract opportunities and experiences that lead to financial success.

It's important to note that the law of attraction is not a scientific theory and the relationship between thoughts and feelings and experiences is not well-understood by science. There is no scientific evidence that supports the idea that we can attract positive experiences and outcomes into our lives by simply focusing our thoughts and feelings on them. However, the concept can be used as a tool to help focus and align our thoughts and feelings with our goals and desires.

In summary, The law of attraction is a popular concept in self-help and personal development circles that suggests that we can attract positive experiences and outcomes into our lives by focusing our thoughts and feelings on them. The concept can be applied in various fields such as career, relationships and financial success. Although there is no scientific evidence that supports the idea, it can be used as a tool to help focus and align our thoughts and feelings with our goals and desires.

In physics, the law of attraction refers to the force of attraction between two bodies due to their masses. This force is known as gravity and is described by Newton's law of universal gravitation, which states that any two bodies in the universe are attracted to each other with a force proportional to the product of their masses and inversely proportional to the square of the distance between them.

One example of the law of attraction in physics can be seen in the orbit of planets around the sun. The sun has a much greater mass than any of the planets, so it exerts a strong gravitational force on them. The planets are attracted to the sun and are kept in orbit around it due to this force of attraction.

Another example can be seen in the behavior of celestial objects such as galaxies and clusters. The law of attraction governs the motion of the objects within these systems, causing them to move towards the center of mass under the influence of gravity.

A third example can be seen in the behavior of objects in space. For example, a satellite that is placed into orbit around the Earth is attracted to the Earth by the force of gravity and is kept in orbit due to the balance of this force with the force of its inertia.

In summary, in physics, the law of attraction refers to the force of attraction between two bodies due to their masses. This force is known as gravity and is described by Newton's law of universal gravitation. This law governs the motion of celestial objects such as planets, galaxies, clusters and satellites and plays a crucial role in the dynamics of the universe.

Entropy is a concept in thermodynamics that describes the disorder or randomness of a system. It is a measure of the amount of energy that is unavailable to do work in a system. The concept of entropy was first introduced by the German physicist Rudolf Clausius in the 1850s as a way to understand the second law of thermodynamics, which states that entropy always increases over time in a closed system.

One example of entropy can be seen in the context of a gas in a container. At the molecular level, the gas particles are in a state of random motion and have a high degree of entropy. As the gas particles collide with the walls of the container, they transfer some of their energy to the walls, reducing their random motion and lowering the entropy of the system.

Another example of entropy can be seen in the context of a chemical reaction. In a chemical reaction, the bonds between atoms are broken and new bonds are formed. This process increases the disorder of the system, increasing the entropy.

A third example of entropy can be seen in the context of a living organism. In a living organism, energy is used to maintain a highly ordered state, such as the organization of cells and tissues. However, as the organism dies, that ordered state begins to break down, and entropy increases.

In summary, entropy is a concept in thermodynamics that describes the disorder or randomness of a system. It is a measure of the amount of energy that is unavailable to do work in a system. The second law of thermodynamics states that entropy always increases over time in a closed system. It can be observed in various fields such as gas in a container, chemical

A self-organizing system is a system that can organize itself without the need for external control or intervention. The organization of the system emerges spontaneously from the interactions between its components. Self-organizing systems can be found in a wide range of fields, including physics, chemistry, biology, and social science.

One example of a self-organizing system can be seen in the behavior of flocks of birds or schools of fish. These animals organize themselves into large, coordinated groups without any central leader or external control. The organization emerges from the simple interactions between the individual animals, such as their tendency to follow the movement of their neighbors.

Another example of a self-organizing system can be seen in the behavior of ants. Ants are able to organize themselves into complex colonies without any central leader or external control. The organization emerges from the simple interactions between the individual ants, such as their tendency to follow the pheromone trails left by their fellow ants.

A third example of self-organizing system can be seen in the behavior of traffic on a highway. Cars on a highway adjust their speeds and positions based on the actions of the other cars around them, without any central control or external intervention. This self-organization leads to the emergence of a pattern of traffic flow that is efficient and relatively stable.

In summary, A self-organizing system is a system that can organize itself without the need for external control or intervention. Self-organizing systems can be found in a wide range of fields, including physics, chemistry, biology, and social science. Examples of self-organizing systems include flocks of birds, schools of fish, ant colonies, traffic on a highway among others. Understanding self-organizing systems can help individuals and organizations to create systems that are more efficient, flexible and adaptive to changing circumstances.

Chaos is a concept in mathematics, physics, and other sciences that refers to the behavior of certain nonlinear systems that appear to be random and unpredictable. Chaos is characterized by sensitive dependence on initial conditions, meaning that small changes in the initial conditions of a system can lead to vastly different outcomes over time.

One example of chaos can be seen in the behavior of weather systems. Weather systems are highly complex and nonlinear, and small variations in temperature, pressure, and other variables can lead to vastly different weather patterns over time. This makes long-term weather forecasting difficult, as small errors in the initial conditions can lead to vastly different predictions.

Another example of chaos can be seen in the behavior of fluid flows, such as in turbulent flows of liquids and gases. Turbulence is characterized by patterns of eddies and vortices that can appear unpredictable and random. The behavior of these flows is highly sensitive to initial conditions and small changes can result in vastly different patterns.

A third example of chaos can be seen in the behavior of certain types of electronic circuits known as chaotic oscillators. These circuits are designed to produce signals that are highly sensitive to small changes in their initial conditions and can appear to be unpredictable and random.

In summary, Chaos is a concept in mathematics, physics, and other sciences that refers to the behavior of certain nonlinear systems that appear to be random and unpredictable. It is characterized by sensitive dependence on initial conditions. Examples of chaos can be seen in the behavior of weather systems, fluid flows and electronic circuits. Understanding chaos can be useful in fields such as weather forecasting, fluid dynamics and circuit design.

In electronics, the law of attraction does not have a specific meaning. The term "law of attraction" is most commonly associated with the concept in self-help and personal development circles that suggests that we can attract positive experiences and outcomes into our lives by focusing our thoughts and feelings on them, as I described in my previous answer.

In electronics, there are various laws and principles that govern the behavior of electronic components and devices. Some examples are Ohm's Law, Kirchhoff's laws, Faraday's law of induction and Lenz's law.

Ohm's law states that the current flowing through a conductor is directly proportional to the voltage applied across it and inversely proportional to its resistance.

Kirchhoff's laws are a set of rules that describe the behavior of currents and voltages in electrical circuits. Kirchhoff's current law states that the total current entering a node in a circuit must equal the total current leaving that node, and Kirchhoff's voltage law states that the sum of the voltages around a closed loop in a circuit must be zero.

Faraday's law of induction states that a changing magnetic field will induce an electromotive force in a conductor that is placed within that field.

Lenz's law states that the direction of an induced current is such that it opposes the change in magnetic field that caused it.

In summary, in electronics, the law of attraction does not have a specific meaning. There are various laws and principles that govern the behavior of electronic components and devices such as Ohm's Law, Kirchhoff's laws, Faraday's law of induction and Lenz's law. These laws describe the behavior of current, voltage, and electromagnetic induction in electronic circuits and devices.

The "give to get" principle is a concept that suggests that in order to receive something, one must first give something of value. It is often used in the context of personal and professional relationships, and is based on the idea that people are more likely to help others who have helped them in the past.

One example of the give to get principle can be seen in the context of business networking. A business person who is looking for potential clients or partners may choose to first offer their own services or resources to others in their network. By doing so, they establish a relationship of trust and goodwill, which makes it more likely that those individuals will be willing to help them in the future.

Another example of the give to get principle can be seen in the context of personal relationships. A person who wants to build a close relationship with someone else may choose to first offer their own time, attention, and support to that person. By doing so, they establish a sense of mutual obligation and trust, which makes it more likely that the other person will be willing to help them in the future.

A third example of the give to get principle can be seen in the context of volunteering. A person who wants to make a positive impact in their community may choose to first give their own time, energy, and resources to a cause or organization that they care about. By doing so, they establish a sense of commitment and investment, which makes it more likely that others will be willing to help them in their efforts.

In summary, the "give to get" principle is a concept that suggests that in order to receive something, one must first give something of value. It can be applied in various fields such as business networking, personal relationships, and volunteering. The principle is based on the idea that people are more likely to help others who have helped them in the past and can be a useful tool for building mutually beneficial relationships.

The phrase "reap what you sow" is a idiomatic expression which means that the consequences of your actions will come back to you. It expresses the idea that the choices and actions that a person makes will have a direct impact on the outcomes and experiences that they have in the future.

One example of reaping what you sow can be seen in the context of personal finance. A person who consistently saves and invests their money will be more likely to have financial security and stability in the future than someone who spends all of their money on unnecessary expenses.

Another example of reaping what you sow can be seen in the context of personal relationships. A person who treats others with kindness and respect is more likely to have positive and healthy relationships than someone who is consistently rude or dismissive.

A third example of reaping what you sow can be seen in the context of career success. A person who works hard and consistently develops their skills and knowledge will be more likely to have a successful and fulfilling career than someone who is lazy or lacks motivation.

In summary, the phrase "reap what you sow" means that the consequences of your actions will come back to you. It expresses the idea that the choices and actions that a person makes will have a direct impact on the outcomes and experiences that they have in the future. Examples of this concept can be seen in personal finance, personal relationships, and career success. Understanding this concept can help individuals make better choices and decisions that will lead to positive outcomes in the future.

Part 4

Applications for your success

So how do you apply these principles to your success?

Well you know everything is ever changing or it becomes unstable and at risk of being chaotic. So you know that things will change and if it is bad it will get better if you let it. If you are successful you know how to manifest the energy of the goal and just let equilibrium do the rest. It makes it easy then to be successful This is how the law of attraction works and it is a universal principle. You know that everything cycles so you know how to take advantage of the cycles to prosper in the market, financially and in life. Buy low sell high is the moto of the stock market so if you can do that repeatedly you can be very well off financially. So to recognize the cycles and indicators is the key to the success. You also know in love that nothing is like the fist kiss. Then one goes into the routine of the relationship. Relationships cycle also. They have their ups and downs.

So understanding the principles can result in you being able to benefit from this knowledge. The key is to recognize the indicators and what is happening to be able to optimize the current point to your benefits.

Best wishes to your success as success breads success like the ripple effect and multiplier effect. It is like money in the bank drawing the compound interest.

Part 5

Further research and observations

Further research and observations can be like how one thing connects to another. For examples is the way the neurons work in our brain the same as how a neural AI network works in artificial intelligence programs? Is the magnitude of the pulse that is transmitted by the neuron the same as the magnitude of the bias in the network allowing it to distinguish between a picture of a dog and cat?

Additional research and observations can be on these principles on how they interconnect in the universe, us and energy. Quantum mechanics seems to follow strange rules could these follow those rules?

The observation and research is your choice as when one things about red sports car all they see is that red sports car everywhere they look. What will you see?

www.ingramcontent.com/pod-product-compliance
Lightning Source LLC
Chambersburg PA
CBHW052359220526
45465CB00003BB/1173